Adam Todd Bruce

Observations on the Embryology of Insects and Arachnids

Adam Todd Bruce

Observations on the Embryology of Insects and Arachnids

ISBN/EAN: 9783742811981

Manufactured in Europe, USA, Canada, Australia, Japa

Cover: Foto ©Klaus-Uwe Gerhardt /pixelio.de

Manufactured and distributed by brebook publishing software
(www.brebook.com)

Adam Todd Bruce

Observations on the Embryology of Insects and Arachnids

OBSERVATIONS

ON THE

EMBRYOLOGY

OF

INSECTS AND ARACHNIDS

BY

ADAM TODD BRUCE,

B. A., PRINCETON COLLEGE, 1881; PH. D., JOHNS HOPKINS UNIVERSITY, 1886.

A MEMORIAL VOLUME.

.

———————

BALTIMORE:
Publication Agency of Johns Hopkins University,
1887.

.

J. D. EHLERS & CO., PRINTERS AND ENGRAVERS,
HOEN BUILDING, BALTIMORE, MD.

This Volume is printed, after the Author's death, as a Memorial, by his instructors, pupils and friends in Princeton, Baltimore and elsewhere. It consists of his Thesis, reproduced from the copy which was accepted by the Board of University Studies of the Johns Hopkins University, at his examination for the degree of Ph. D., in April, 1886.

THE

SCIENTIFIC WORK
OF
ADAM TODD BRUCE.

A SKETCH
BY
W. K. BROOKS.

ASSOCIATE PROFESSOR OF MORPHOLOGY, JOHNS HOPKINS UNIVERSITY.

Adam Todd Bruce was born March 6th, 1860; he graduated at the University of New Jersey in 1881; he obtained the degree of Ph. D., at the Johns Hopkins University in June, 1886; he was appointed an Instructor in the Johns Hopkins University in September, 1886, and his death took place in March, 1887.

During the three years of my acquaintance with Dr. Bruce, the rapid development of his strong character was a constant pleasure to me, but while each day added strength to my belief that a most useful and distinguished career lay before him as a scientific investigator, I sorrow for the loss of an affectionate friend, rather than the death of a bright and promising pupil and student of nature.

Our relation was not that of instructor and pupil, but an intimate personal friendship; to me he had brought his pleasures and successes, and his troubles and perplexities, and when I received, by the first mail which reached me at the Bahama Islands, the sad news of his death, I keenly felt the isolation which added intensity to the loss, and cut me off from a share in the expression of affection for his memory by his friends in Baltimore.

As I was not able to attend the meeting of his friends in Baltimore, when the news of his death was received, it is a great pleasure to have this opportunity to speak of the value of his scientific researches, and of the great loss which science has sustained by his untimely death, just as he was beginning to apply his skill and his thorough training in technical methods to the solution of new problems.

As a student and investigator Dr. Bruce was eminently characterized by perseverance. Always alive to all that was passing around him, he was keenly interested in many things, but no outer attraction could draw him from the subject to which he had set his hand. Difficulties only added zest to the work, and his indefatigable zeal never flagged, even when there seemed to be little hope of success.

All the members of our party, at Beaufort in 1885, will remember his efforts to encourage out-door exercise in a climate where exercise is so repulsive. As I watched, at evening, his boyish enthusiasm in athletic games, I could scarcely believe that he was the same person who had struggled all day long, week after week, with the perplexing obstacles which an unfavorable climate, the absence of all adequate facilities, and the most refractory character of his material opposed to his study of the Embryology of Limulus.

In the laboratory all his strength of mind and body was bent upon the problem in hand, but I always felt the fullest assurance that he was in no danger of becoming a narrow specialist, for whenever he was not employed in research, or in the recreations into which he threw himself with equal energy and enthusiasm, his mind was occupied with the consideration of the bearing of his special researches, both upon the broad problems of humanity, and upon his own personal training and development.

As regards his scientific researches, I have to speak rather of his promise than of results achieved.

Cut off as he was, just as he was beginning to show his power to question nature for answers to new problems, most of his results were very incomplete, and as he had worked, here and there, upon a very comprehensive subject, the embryology of Arthropods, upon which he had purposed to spend several years, he had done little or nothing to bring together his various lines of research.

When he presented himself for examination for the degree of Ph. D. he had made many observations upon the embryology of Insects and Arachnids, and although he was still carrying on this research, and adding daily to his store of information, he, at my advice, wrote out, and submitted as his thesis, a statement of the results of his study up to that time.

During the following year he continued the work, and if he had been able to re-write the thesis, he would have made many important additions. As he was, by nature, incapable of dwelling upon a subject after he had reached his results, he

hurried on to new fields, and made no drawings to illustrate his conclusions. As embryological notes must be in pictorial form, most of his work is unrecorded and as his thesis is his single complete and illustrated work, it has seemed proper to his instructors, pupils and friends, to publish it. Although he himself would have wished to include his later results, this is now impossible, and it is here printed exactly as he wrote it in the winter of 1875-6.

In the summer of 1876 he made, conjointly with myself, a study of the embryology of Limulus. This was very complete as regards the early stages, and we had intended before publishing, to spend one or two years more upon it, and to make it an exhaustive account of the entire embryology and anatomy of Limulus, but the publication of Kingsley's fully illustrated paper upon the later stages of development, and my own employment with other subjects, and Dr. Bruce's desire to study other forms for comparison, retarded the completion of the work.

The segmentation of the egg, the formation of the blastoderm and of the germ-layers, and the anatomy of the young larva, were all thoroughly studied, and illustrated by nearly a hundred drawings, and I hope that some means of publication will soon be found.

Upon these two papers Bruce's reputation among strangers, as a student of nature, must rest, but all who knew him personally will feel how very inadequately this will represent what a few years more might have produced.

As soon as the news of Bruce's death was received in Baltimore a large number of his friends and pupils met in the biological lecture room of the University to take action indicating their grief at his death and their sympathy with his relatives.

On motion of Dr. Howell, Professor Martin took the chair; he requested Mr. Washburn to act as secretary.

Professor Martin said: "My friends, the University has not existed for its brief ten years, without the hand of death showing its power among us. Those who have gone have been of almost every degree of academic dignity, from professor to undergraduate. We sorrow when the old man dies, but we know that he has lived to do his work and make his name and fame. We grieve when the bright lad dies, but we know that he has been spared years of toil and struggle. Surely no death is so sad as that of a young man, who has just completed seven or eight years of hard work at college and university, and is beginning to enjoy the fruits of his labors. Such was the death which is the occasion of this meeting.

"Adam T. Bruce graduated at Princeton in 1881, and remained there for two years after, as graduate student and instructor. He was certainly a member of one, and I believe of two, of the paleontological expeditions to the far West, through which Princeton men have done so much for science. Bruce was elected a Fellow of this University in 1883, and appointed a Fellow by Courtesy in 1885. He took his degree as Doctor of Philosophy last June.

"Though devoted to morphological work he was not narrow in his sympathies or pursuits. He had a great fondness for English literature, especially the older literature, and had a very extensive knowledge of it. At Princeton, his studies were largely philosophical while he was an undergraduate, and after coming here he did a considerable amount of psychological work under the direction of Professor Stanley Hall. From his first appearance among us he was an active and valuable member of the foot-ball team.

"His handsome vigorous frame, his bright pleasant face, his manly honest look, made all who met inclined at once to like him : and those who knew him, esteemed him more, the more they knew him. So that between those who loved him for himself, and those who esteemed him for his work, and those who were his comrades in athletics, he had among us a very large number of friends, representing many departments and many interests in the University. Outside the University he had many friends in Baltimore, won by his bright intellect and gentlemanly bearing.

"During the last summer vacation, Bruce was offered the post of Instructor in Osteology and Mammalian Anatomy. In accepting, he wrote to me expressing the great pleasure it gave him to continue his connection with the University. Those who were his pupils know with what energy and enthusiasm he entered on his work. But, great as his strength was, he overtasked it. The previous spring he had been reading hard for his degree examination ; he had been at Wood's Holl at work most of the summer ; and the completion of his research on Limulus, together with his lectures, was too much for him. Towards the close of November he found himself obliged to give up all work for a time, and started on an extended tour with the aim of resting and recuperating, intending finally to stay some time in Japan and study some of the Pacific forms of life. While passing through Egypt he was seized with a fever—and the end came.

"It seemed to some of us, that all who knew him here would be glad of an opportunity to meet and testify to our affection for his memory and our sorrow for

his death. For that reason this meeting has been called."

Dr. A. L. Kimball, who had been a classmate of Dr. Bruce at Princeton, then said a few words in regard to the esteem in which Dr. Bruce had there been held, and proposed the adoption by the meeting of the following preamble and resolution:

Whereas, We have learned with profound sorrow of the death of Dr. Adam T. Bruce, the friend of all, and the instructor of many of us—and

Whereas, He had especially endeared himself to us by his unfailing kindness and courtesy—it is

Resolved, That, assembled here to-day, we love the memory of the pleasant companionship which existed between him and us, in all relations, official and personal; and express our grief that he was not spared to return among us—and

Resolved, That we tender to his immediate friends and relatives our sincere sympathy in their bereavement—and

Resolved, That a copy of these resolutions be forwarded to the editor of the University Circulars, to the Faculty of Princeton College, and to the Family of Dr. Bruce.

Mr. Coleman, one of Dr. Bruce's pupils, seconded the motion of Dr. Kimball, and the following gentlemen spoke of the high regard and affection with which they remembered him, viz: Mr. Coleman, Mr. Reese, Mr. Riggs, Professor Hall, Dr. Hartwell, Mr. Washburn, Mr. Gilpin, Mr. Burton, Dr. Jastrow, and President Gilman.

The Preamble and resolutions proposed by Dr. Kimball were then unanimously adopted, and the meeting adjourned.

H. NEWELL MARTIN, *Chairman.*

F. L. WASHBURN, *Secretary.*

It only remains for me to signify on my own part, and in behalf of the students who were with me at Nassau, our regret that we were unable to take part in this meeting, and our concurrence with its action, and to express our affection for the memory of our lost friend Adam Todd Bruce.

OBSERVATIONS ON THE EMBRYOLOGY

OF

INSECTS AND ARACHNIDS.

The important points to be determined in Insect embryology are, the segementation of the egg and formation of the blastoderm—the origin of the embryo and embryonic membranes—the formation of the germinal layers—metameric segementation and all connected with it—including number of appendages, nerve ganglia and etc.

The embryology of Arachnids, or at least of spiders, in some important particulars resembles the embryology of Insects. These points of likeness will be brought out in the detailed description of the embryology of spiders which will be given later.

The Insects studied included representatives from the Lepidoptera, Coleoptera and Orthoptera; while a few incomplete observations were made on the embryology of the Neuroptera and on the maturation of the ovum in Musca. The ova were studied by means of sections. With opaque eggs the sectional method of study certainly throws more light on important embryological points than mere superficial observation.

The eggs were hardened in corrosive sublimate and in Perenyi's fluid. The chorion was punctured to admit hardening and staining fluids. After staining they were passed through the various grades of alcohol and were then transferred to chloroform—they were then placed in a mixture of chloroform and parraffine, in which they remained a shorter time. They were then embedded in paraffine and were readily cut to any degree of thinness required

A detailed description of each species studied will be given, commencing, not with the lower insect groups and spiders, but with the highest group studied, viz: the Lepidoptera.

General conclusions and comparisons will be reserved for the closing paragraphs of this paper.

LEPIDOPTERA.

The Lepidoptera were represented in this study by *Thyridopteryx ephemeraiformis* or the common bag worm.

The observations made on the embryology of this insect owing chiefly to abundance of material were more complete than those made on any other insect studied by the writer. Consequently the description of its development is given first that it may serve to fill out as far as possible the incomplete accounts given of the development of other orders of insects. The development of Thyridopteryx from the earlier stage when all the important organs of the embryo had been formed was carefully followed. The study of the embryology of this insect, if it has brought to light nothing new, has, in the opinion of the writer at least, settled some important points connected with the embryology of this group of insects.

The manner in which Thryridopteryx, commonly known as the bag worm, lays its eggs, has been recently described by Professor Riley. A female kept in confinement left the cocoon after laying her eggs in it, crawled a short distance and died. Professor Riley states that in a state of nature the female leaves the cocoon, falls to the ground and dies. The unsegmented ovarian egg of Thridopteryx (Fig. I & I') consists of an outer protoplasmic stratum shading off into the central yolk spherules which are probably invested with protoplasmic ramifications extending from the outer stratum.

Referring to figure I, which represents a highly magnified portion of the section represented by figure I, it will be seen that the egg is invested by a thick chorion (CH in Fig). The chorion is perforated at the pole of the egg by a conspicuous micropyle which is not represented in the figure.

Outside the chorion the ovarian epithelium (OE in Fig.) is seen. It consists of flattened cells with large oval granular nuclei. The granular protoplasmic stratum of the egg (P, in Figs.) lies beneath the chorion and extends between the yolk spherules. The spherules (YS in Fig.) are frequently vaccuolated. No trace of a germinal spot or vesicle was found in the mature egg. The former might readily have been present and have escaped observation.

The early stages of segmentation have already been studied in the Lepidoptera by Bobretzky.[1] In the earliest stages he found four amoebiform cells in the yolk situated in pairs at opposite ends of the egg. Later the blastoderm is formed by a multiplication of these cells: according to Bobretzky it is not formed simultaneously over the entire surface of the egg, but is laid down first at one or more points on the surface. This type of segmentation cannot strictly be called entolecithal, in as much as the cells are not, in the earliest stages of segmentation, at the surface enclosing the yolk. All the primitive undifferentiated cells do not, according to Bobretzky, reach the surface to form the blastoderm, but some remain centrally located as yolk cells after the formation of the blastoderm. The earliest stages of segmentation observed in Thyridopteryx showed several amoebiform cells in the yolk in each cross section.

The egg at this stage must consequently contain a number of cells most of which have not reached the surface. The nuclei of these centrally located cells generally have indistinct boundaries which become more distinct and their contents clearer on reaching the surface (Fig. III'). Figure III' represents a cell at the surface undergoing division. It has two nuclei with clear contents and well defined boundaries.

The other cell represented in the figure has just reached the surface of the egg. Its nucleus is indistinct and granular. Much reliance cannot be placed on histological details like these, however, except in case of remarkably well preserved specimens.

In Thyridopteryx, as in the Lepidopterous insects studied by Bobretzky, the blastoderm does not appear simultaneously over the entire surface of the egg but appears first at one point on the surface. Figure III is a transverse section of the egg at this stage. The yolk spherules are not represented in the figure. It will be observed that the blastoderm is only formed over a portion of the surface.

Closely related insects are said to differ in their blastoderm formation, however, and the simultaneous or unsimultaneous appearance of the blastoderm over

[1] Bobretzky. Uber die Bildung des Blastoderms und den Keimblatter bei den Insecten.

the entire surface is consequently a fact of little embryological importance.

The first trace of the embryo is a thickening of the blastoderm at one point where it becomes more than one cell thick (Figs. IV & IV'). Whether this thickening results from the division in a single layered blastoderm at that point, or from a special aggregate of embryonic cells occurring there, could not be satisfactorily determined. Figure IV represents a transverse section of an egg showing the early embryo or ventral plate, E in fig. in cross section. Figure IV" is a drawing of the embryo at this stage more highly magnified. On either side of the embryo, as seen in cross section, are seen folds of blastoderm (Am F. Fig. IV') which ultimately grow together and unite over the median line of the embryo. On the union of these folds, the embryo with the inner limb of the fold or true amnion becomes separated from the outer fold or serosa, and comes to lie in the yolk (Fig. V), while the serosa remains continuous with the blastoderm. At the stage represented by Figure IV it should be noted that some cells have not reached the surface, but remain in the yolk enclosed by the blastoderm (YC Fig. IV'). These are the so called yolk cells. Concerning the origin and fate of these cells in different insect groups there is considerable difference of opinion among investigators. Some investigators, notably Balfour, in his work on Embryology, from evidence afforded by Bobretzky's work on the Lepidoptera just referred to, and from other observations, claims that these central yolk cells are in all cases undifferentiated cells which have never reached the surface, and homologizes them with the endoderm of Arachnids. On the other hand, Patten [1] and probably Graber [2] believe that a stage occurs in the development of the insect egg in which there is a blastoderm surrounding a central yolk-mass containing no cells.

The yolk cells which are present in the yolk at subsequent stages must migrate from the blastoderm if this view be correct. Probably both views are correct. In the Lepidoptera from Bobretzky's observations, and from the study of Thyridopteryx, there seems to be little doubt that the first view is true. But from a study of some other groups of insects it seems probable that the second view is also true of those groups.

To return to the embryo, which it will be remembered had been separated from

[1] Patten. Development of Phryganids. Quarterly Journal of Microscopical Science, 1885.
[2] Graber. Archiv. f. Mikr. Anat. Vol. XV.

the surface by the union of the amnion folds, and had come to lie in the yolk: Figure V represents a transverse section of the egg at this stage with the embryo (E in Fig.) lying in the centre of the yolk. The embryo at this stage, as shown by studying a series of sections, is watch-glass shaped, with its convex ventral side covered by the amnion, which it will be remembered is the inner limb of the folds described, and is constricted off from the surface of the egg with the embryo. In Figure V the amnion covering the ventral surface of the embryo is not represented, but on referring to Figure VII, which is a cross section of an older embryo, the structure and relation of the amnion (AM in Fig. VII) covering the ventral surface of the embryo will be understood. The early watch-glass shaped embryo (Fig. V) consists of a single layer of elongated epithelial cells with nuclei in some cases in process of division on their ventral extremities, i.e. on the extremities at the convex or ventral side of the embryo. A few nuclei with aggregated protoplasm about them lie on the concave side of the embryo (YC Fig. V). These nuclei with the protoplasm about them are probably cells about to migrate from the embryo into the yolk.

In the earliest stages it will be remembered that the embryo was more than one cell thick in some places (Fig. IV). Probably some of the cells of this early embryo remaining on its dorsal or concave surface are the cells just described as about to migrate from it into the yolk. Before the embryo reaches the stage represented in Figure V the yolk undergoes segmentation. The segmentation of the yolk is probably due to the multiplication of yolk cells. Radial processes from the protoplasm of these cells adhere to a number of yolk spherules. Consequently there is formed a structure known as the yolk ball or yolk segment, consisting of a central cell, radial protoplasmic processes of which hold yolk spherules in position. YB Figure XI represent a cross section of a yolk ball. Similar cross sections will be observed in other figures. Adjacent yolk balls appear in some cases to be united; some are connected with the blastoderm and others with the amnion or embryo, consequently the entire egg is probably a protoplasmic continuum with all its parts in connection. The structure of the blastoderm cells is shown in Figure VII. They are flat hexagonal cells with conspicuous nuclei. On the union of the amniotic folds, the embryo with the true amnion is constricted off from the surface of the egg as described. With the sinking of the embryo into the centre of the egg the yolk comes to lie between the amnion and blastoderm. In the space between the amnion and ventral surface of the embryo, however, there is no yolk. The amnion consists of flat hexagonal cells resembling the blastoderm cells in appearance. Later the embryo lying in the centre of the yolk becomes pear-shaped, with its broad anterior end deeply cleft. The shape of the embryo at this stage will be understood from an examination of Figs. VIII, VIII' & IX. Figure VIII, shows the outline of the embryo at this stage as indicated by a horizontal section. Figure VIII' represents diagrammatically a longitudinal section of an embryo at about the same stage of development. The lateral portions of the broad anterior end of the embryo, separated by the deep cleft are the so called pro-cephalic lobes (P C Fig. VIII). Figure IX is a less diagrammatic reproduction of a horizontal section of an embryo at the same stage of development as figure VIII. B L, in Figure IX stands for the blastoderm. The space between the blastoderm and embryo in the section, was of course filled with yolk balls, only one of which (Y B in Fig. IX) is drawn.

The cleft separating the pro-cephalic lobes is a continuation of a median groove extending along the median line of the ventral surface of the embryo at this stage. This groove or depression on the ventral surface of the embryo causes an elevation on its other surface, that is, on its surface towards the future dorsal wall which has not yet been formed. Figure VII represents a highly magnified transverse section of an embryo at a stage when the median groove deepens, beginning to push dorsally the median portion of the embryo (D in Fig. VII). In subsequent stages the groove deepens, and the pushed in portion of the embryo becomes folded off, and forms the inner layer (i in Figs). The inner layer is not strictly invaginated, for it is cut off from the rest of the embryo before the opposite sides of the median groove have met. After its separation the inner layer splits into two bands.—The line of separation of these two bands being immediately beneath the blastopore. Figure X is a transverse section of an embryo at this stage. The inner layer (I in Fig. X), it will be observed, is divided into bands, the line of separation being a continuation of the median groove (BH in Fig). Figure X' represents the same section as X drawn with a lower power, showing the relation of the embryo to the yolk balls and blastoderm (BL.). On its separation into two bands or shortly after the separation has been effected, each band of the inner layer segments or becomes divided into a

series of somites. The somites thus formed are solid. Figure VIII' is a diagrammatic drawing of a longitudinal section of an embryo at this stage of development.
The somites of the inner layer are seen in longitudinal section (DS, in Fig. VIII).
It will also be seen that the body somites are marked by incisions of the outer layer
or ectoderm extending between the somites of the inner layer. The inner layer extends the entire length of the embryo.

The dotted line joining the ends of Fig. VIII' represents the amnion in longitudinal section. The extreme ends of the embryo, in referring to figure VIII', are
seen to be curved dorsally, that is, away from the amniotic side of the embryo. The
two terminal somites thus have their body wall formed dorsally before the dorsal
surface of the intermediate portion of the embryo is closed.

At a later stage of development the ectoderm becomes thickened on each side
of the blastopore; consequently there are formed two thickened strings of ectoderm
which are the first indications of the nervous system. Figure XI is a drawing of a
transverse section of an embryo at this stage. NS in Figure XI represents the
thickened ectoderm lying on each side of the blastopore (BH in Fig.).

Subsequently the thickened ectoderm strings which are to form the nerve cords
become separated from the superficial ectoderm, and also from the ectoderm covering
the sides and bottom of the blastopore. The ectoderm cells at the bottom of the
grove or blastopore (BH, Fig. XII) become greatly elongated. These elongated
cells lie close to the nerve cords, but do not actually join them. Figure XII, represents a transverse section of an embryo at this stage of development. The nerve
cords (NS in Fig.) will be seen to be separated from the surface ectoderm lining the
groove. The elongated ectoderm or what might more properly be called indifferent
cells at the bottom of the groove will also be observed in the figure. When the
nerve cords have been separated from the surface ectoderm they segment;—the number of somites corresponding to the number of body segments.

Hatschek, [1] from his studies on the Lepidoptera, thought the elongated ectoderm or indifferent cells lying at the bottom of the blastopore, ultimately fused with
the nerve ganglia and formed the long commissure of the nervous system.

In this, as subsequent stages in the development of Thyridopteryx show, he

[1] Beitrage. Zur Entwicklung d. Lepidoptera. Jenaische Zeitschrift Bd XI.

was probably mistaken: for the elongated cells in question divide in subsequent stages, and thus give rise to cells corresponding in all particulars to migratory mesoderm cells (I' Figs. XII & XV). All the stages in the division of these cells and their conversion into migratory mesoderm were not traced, but after a study of the sections drawn, and from many others not drawn, there can be little doubt that the cell mass at the bottom of the blastopore takes no part in the formation of nervous tissue. The non-participation of these cells in the formation of nervous tissue is also shown from a study of the development of the nerve commissures.

In Figure XV, which represents a transverse section of a well advanced embryo, it will be seen that a granular substance, probably formed by the breaking down of cells, has appeared on the dorsal surface of the ganglia (NS in Fig. XV). The cells at the bottom of the blastopore take no part in the formation of this granular substance. Still later in embryonic development the commissures, both longitudinal and transverse, can be recognized as extensions of the granular material formed near the dorsal surface of the nerve ganglia. In these later stages there is no trace of elongated cells at the bottom of the grove, but between the nerve ganglia and forming the peritoneal coat or neurilemma of the nervous system are migratory mesoderm cells.

Figure XXIII represents two thoracic and the first two abdominal ganglia in median longitudinal section. Between the ganglia will be seen the migratory mesoderm cells (I') which probably arise in the manner just described, forming the neurilemma or peritoneal coat of the nervous system, but evidently taking no part in the formation of commissures. Only a portion of the migratory mesoderm cells arise from cells lying between the nerve cords: the greater portion of it is probably derived from cells of the inner layer. These migratory mesoderm cells have a round cell body which does not stain, containing a nucleus which stains deeply. They are similar in appearance, though apparently not in origin, to migratory mesoderm cells described by Reichenbach for the Crustacea.

The supra-oesophageal ganglion arises differently from the other nerve ganglia. It appears first as a thickening of the lateral portions of the procephalic lobes. Figure XIII is a drawing of a section through the procephalic lobes; the lateral portions of which (I in Fig.) are thickened considerably. On the median portion of the procephalic lobe (LB in Fig.) is the paired outgrowth which forms the upper lip. The dorsal surface has been partly covered by the amnion (AM' in Fig.) Yolk spherules are seen filling the body cavity.

Figure XIII' is a drawing of a section through the procephalic lobes of a more advanced embryo. Here it will be observed that the inner cells of the lateral portions of the procephalic lobes (I in Fig.) have become specialized as nerve cells. Later these specialized cells are separated from the outer unspecialized ectoderm, and form on each side the halves of the supra-œsophageal ganglion. Figure XXII represents a good longitudinal section of the supra and subœsophageal ganglia.

The Supra-œsophageal ganglion consists of two portions; an anterior portion (I in Fig.) innervating the antennæ, which are not represented in the figure, and a posterior portion (No. I' in Fig.) which sends a nerve (LBN in Fig.) to the labrum or paired lip.

Posteriorly the second division (No. I' in Fig.) of the supra-œsophageal ganglion forms part of the circum-œsophageal commissure, which is completed by a portion of the mandibular division of the sub-œsophageal ganglion, (No. II in Fig).

The supra-œsophageal ganglion has its opposite halves united by two commissures; an anterior commissure (C Fig. XXIX Plate 3) extending beneath the œsophagus, and a posterior commissure (C" Fig. XXXI) extending above the œsophagus.

The sub-œsophageal ganglion consists of three pairs of closely connected ganglia innervating the mandible, first, and second maxillæ, respectively (Nos. I, II, & III Fig. XXII). The three pairs of thoracic ganglia are larger than the following ten pairs of abdominal ganglia. This difference in size is apparent on consulting figures XVII, XVIII, XIX, which represent good longitudinal sections of an advanced embryo. There are ten pairs of abdominal ganglia, the last or tenth pair being smaller than the preceeding (No. 17, Fig. XVII). It is questionable whether the terminal portion of the abdomen which forms the so called eleventh abdominal somite is to be regarded as a true somite or not. It has no ganglion corresponding to it, and is formed, as stated, by the dorsal flexure of the posterior end of the embryo, and consequently has its body walls formed on all sides at an early period in the manner described.

The nervous substance consists principally of round nuclei. The commissures consist of the granular material before described. The nerve fibres consist of similar granular material (Figs. XXII & XXIII). After the separation of the nerve ganglia from the ectoderm, the blastopore closes entirely (Fig. XV). A closer examination of figures XVII, XVIII, XIX, may make clearer the number and relations of the nerve ganglia.

Figure XVII is a nearly true median longitudinal section The œsophageal invagination (OE in Fig) will be seen directly posterior to the upper lip (LB in Fig). Following the three large thoracic ganglia are the ten smaller abdominal ganglia, (Nos. 8-12, Fig). The last abdominal ganglion will be seen to be smaller than the preceeding ganglia.

The nature and origin of the last or eleventh abdominal somite has already been described. Its dorsal wall extends forward beyond its lateral walls. This dorsal extension of the last somite will be seen, on referring to figures XVII, XVIII, & XIX, to be deeply grooved. This groove is a continuation of the median groove or blastopore over the dorsal surface. It becomes deeper, and is finally invaginated. On being separated from the dorsal surface of the body, the invaginated portion forms the anal invagination. The œsophageal invagination also occurs in the middle line of the body; unlike the anal invagination, however, it is formed not by the infolding of a portion of the median groove, but by a simple, vertical ingrowth (OE Fig. XVII). The œphagus is formed before the anal invagination is completed.

Back of the œsophagus there occurs an aggregate of migratory mesoderm cells, wrongly regarded by Hatscheck as endoderm. [1] The œsophageal invagination occuring in the median line of the body pushes in the cells of the inner layer at that point, and extends dorsally and posteriorly, invested by these cells, (Figs. XVIII, XXII). On the dorsal surface of the œsophageal invagination there is formed a thick string of mesoderm cells. The outer cells of this string adhere together, while the inner cells separate from them; thus is formed a tube containing cells. This tube is the heart (H in Figs. XXVIII Plate 2, & XXIX, XXX, XXXI Plate 3).

Probably the inner cells of the tube form the blood corpuscles and plasma.

Beyond the œsophageal invagination the heart does not appear to be formed from a solid string of cells, but is apparently formed from separate mesoderm cells which come together on the dorsal surface, over the alimentary tract, flatten out, and adhere at their edges (Fig. XXXI). A more detailed description of figures XXVII, XXXIII will make preceeding references clearer.

Figures XXVII & XXVIII represent sections through the head. The head ganglia are not observed in these figures, because the sections were made through the ventral surface of the head outside the nervous substance.

The œsophagus (OE in Figs.) is represented in cross section, as well as the mandibles (MD) and labrum (LB).

The œsophageal epithelium is represented as surrounded by mesoderm carried in with it as it grew inwards from the median line. The mesoderm surrounding the œsophageal ingrowth, on its dorsal surface, will be seen to be hollow, thus forming what appears to be the first trace of the heart in the head region (H in Figs).

A line joining numbers I and II in figure XVII, will represent approximately the plane of the sections represented by figures XXVII and XXVIII. Succeeding sections of the same series represented by figures XXIX, XXX and XXXI, were cut in planes parallel and internal to the imaginary line joining numbers I and II in figure XVII. In figure XXIX portions of the supra, and sub-œsophageal ganglia are represented (Nos. I and II in Fig.) The sub-œsophageal commissure of the supra-œsophageal ganglion is also represented (C in Fig. XXIX.)

The structure marked GL in the figures is a prolongation of the salivary gland which originates at the base of the mandibles (GL, Fig. XXVIII). Figure XXX represents a section through a plane internal to that represented by the preceeding figure. The section figured has been carried through the base of the brain, cutting the circum-œsophageal commissure (C' in Fig.)

Figure XXXI represents a section internal or dorsal to the preceeding, showing the supra-œsophageal commissure of the supra-œsophageal ganglion (C" in Fig.)

Figure XXXII is a drawing of a cross section of the embryo back of the œsophageal ingrowth. The epithelium of the midgut (IE in Fig.) is represented enclosing yolk cells and yolk spherules (YS, YC in Fig.)

It may be here in place to describe the sense organs and appendages of the head. The head which includes the portion of body containing the supra and sub-œsophageal ganglia is distinctly separated from the thorax.

Figures XXIX' and XXX' Plate II are diagrammatic representations of the head viewed from its under surface and laterally. The upper lip is seen to be distinctly bilobed. The first maxillæ are longer than the other cephalic appendages.

The sense organs are what may be termed compound simple eyes. They are represented in figures XXVII and XXVIII. Each ocellus, (OC in figures,) apparently arises as an involution of cells from the inner surface of the ectoderm. These involutions form sacks apparently containing no cells; but the lumen or cavity of the sack is filled with granular material. Migratory mesoderm cells probably invest these ocelli.

It does not appear that any part of the ocelli is formed from the nervous system. The observations which were made were not conclusive on this point however.

An invagination (GL, Figs. XXVII & XXVIII) occurs at the base of the mandible which apparently forms the salivary gland.

It remains now to describe the origin of the endoderm and the closure of the dorsal surface of the body. The appendages and tracheal invaginations arise in the customary manner.

A portion of the inner layer on each side of the embryo becomes separated from other parts of the inner layer (IE Figs. XV, XXIV, XXV.)

These portions of the inner layer which may be called endoderm grow together and unite first on what is the ventral surface of the alimentary tract (Fig. XXV.) They then extend dorsally and enclose the yolk and the yolk cells which lie in the body cavity.

Figure XXXII is a cross section, already explained, of an advanced embryo in which the epithelium of the midgut is fully formed. The yolk cells lie, with the yolk, in the digestive tract and certainly do not form any considerable portion, if any, of the endoderm. A similar formation of endoderm has been described by Tichomiroff[1] for other Lepidoptera. His conclusions have been disputed, notably by Balfour, who claims that the yolk cells are the true endoderm. Before its closure by the endoderm, the yolk is enclosed by migratory mesoderm cells (I' Fig. XXIV.) Inside these cells the endoderm grows round and encloses the yolk.

(1) Zool. Anzeiger. II Jahr. no. 20.

The migratory mesoderm cells surrounding the digestive tract may have been wrongly regarded by Hatscheck as endoderm since, as already stated, he seems to have mistaken other migratory cells for endoderm.

The closure of the dorsal surface of the body is interesting. The amniotic folds (AM' Fig. XXV) grow dorsally more rapidly than the ectoderm of the body walls. These opposite amniotic folds finally unite and the inner limb of each fold forms a portion of the dorsal surface of the body. The outer limbs of the amniotic folds unite to form a sack in which the entire embryo lies. This sack contains no yolk, but is apparently filled with fluid. Just before the union of the opposite amniotic folds, there is formed what may be termed a dorsal organ (Fig. XXVI,) though it does not correspond to the dorsal organ described for some other insects.[1]

A study of figures XV, XXIV, XXV, XXVI, will make clearer what has been said concerning the origin of the endoderm and the union of the amniotic folds on the dorsal surface of the body.

Figure XV has already been described. It represents a transverse section of a well advanced embryo. The tracheal invaginations are shown in the figure (TR) to be invaginations of ectoderm. The appendages are outgrowths of the body cavity occurring between the tracheal invaginations and the nerve ganglia.

On comparing figure XV with figure XI or with preceeding figures it will be seen that in the former, which represents a transverse section of a more advanced embryo than the latter, the amniotic folds have extended farther towards the dorsal surface of the figure. Figures XXIV, XXV, XXVI, are drawings of an older embryo than that represented by figure XV. Here it will be seen that the inner portions of the amniotic folds (AM' in Figs.) are growing together and forming the dorsal wall of the body as described.

Figure XXVI represents a section of an advanced embryo just before the union of corresponding portions of the amniotic folds of opposite sides, when the outer and inner portions of each fold are united by intermediate amnion which forms a sort or dorsal organ.

Inasmuch as no yolk lies between the embryo and the amnion it will readily be

---

[1] Brandt. Beitrage zur Entwicklungsgeschichte d. Libellula.

seen on referring to the figures that on the union of the outer portions of the opposite amniotic folds the sack enclosing the embryo will contain no yolk.

How it happens that folds are formed by the dorsal growth of the amnion will be understood by referring to figure XV. In describing the origin of the endoderm reference was made to figures XV, XXIV, XXV, as illustrating stages in the formation of the midgut.

In figure XV the most dorsal portion of the inner layer (IE in Fig.) on opposite sides begins to be constricted off from the rest of the inner layer.

The process of constriction has been completed in figure XXIV. The portions of the inner layer thus constricted off (IE Figs. XXV, XXVI), on each side grow together forming first the ventral surface of the digestive tract, thence they extend dorsally, and shut in the digestive tract on all sides.

To sum up the embryology of Thyridopteryx: It was found that on the formation of the blastoderm some of the cells, probably, do not reach the surface, but remain in the yolk as yolk cells, which take little, if any, part in the subsequent formation of endoderm.

The nervous system arises in the customary manner. The median ingrowth between the nerve ganglia takes no part, as Hatscheck thought, in the formation of the commissures. The supra-œsophageal ganglion is double. It has a double commissure uniting its opposite halves.

The ocelli probably arise from ectoderm independently of the nervous system. The true amnion on the union of its folds forms a portion of the dorsal surface of the body. Before the union of its opposite folds what might be described as a dorsal organ is formed.

NEUROPTERA.

Chrysopa was the representative of this group which I studied. The observations made on the embryology of this insect, owing to lack of material, were very incomplete.

Figure XXXIII represents a transverse section of the head showing the position of the upper lip and the antennæ. The upper lip is distinctly bilobed. Its two lobes are not, however, well shown in the figure.

The abdomen of the neuroptera, according to Packard, consists of eleven somites including the last somite or the so called post abdomen. The number of somites is not well shown in figure XXXIV, which is not strictly median. The last or eleventh abdominal somite is covered dorsally and corresponds to the last somite of Thyridopteryx. According to Packard, the last somite, or post abdomen, consists in Diplax, [1] of several somites.

This segmentation of the post abdomen was not observed in Chrysopa through the observations made on this insect were confessedly very incomplete.

COLEOPTERA.

The embryology of Meloe, the parasitic beetle, was studied as representing the Coleoptera. The earliest stage obtained showed a surface blastoderm enclosing central cells (Fig. XXXV). Later the blastoderm cells become more columnar, (Fig. XXXV') and there are apparently no cells in the yolk; though unless a complete series of sections of a very well preserved egg be obtained, it is impossible to be positive on the latter part. Apparently, however, a stage occurs in which there are no cells in the yolk, consequently the yolk cells, numerous in latter stages, probably arise from the blastoderm or from the embryo. The earliest embryo (Fig. XXXVI) showed, in cross sections, incipient amniotic folds on each side with a median blastopore.

It becomes separated from the surface on the union of the amniotic folds, but does not sink into the centre of the egg as does the embryo of Thyridopteryx, (Figs. XXXVII, XXXVIII).

Amœbiform cells (YC' in Figs.) can be seen on the dorsal or yolk side of the embryo at this stage. These are probably yolk cells which arise by division from the embryo and subsequently migrate into the yolk. The median invagination becomes constricted of and forms the inner layer, (Fig. XXXIX).

No observations were made on the origin of the nervous system or on the origin of the endoderm. The nerve ganglia become separated from the surface ectoderm, and in advanced embryos all trace of the blastopore has disappeared as in Thyridopteryx (Fig. XL).

[1] Packard. Guide to the study of Insects pp. 56-57.

The upper lip is double and the antennæ have the same position as in Thyri-dopteryx and Chrysopa.

ORTHOPTERA.

Mantis and the grasshopper were the orthopterous insects studied.

The early stages in the development of Mantis were obtained and the latter stages in the development of the grasshopper. The developmental history of the group derived from the study of both insects is not complete, though some observations of importance were made.

At the earliest stage obtained, the egg of the grasshopper consisted of large angular yolk masses like those of the spider (Fig. XLV). The yolk is enclosed by a very thick membrane with concave hexagonal depressions on its outer surface (CH, Fig. XLV, XLVI). Within this outer chorion there is another thinner mem-. brane.

In what is apparently the earliest stage, a portion of the yolk is much vesiculated (E' Fig. XLV). Near or in this portion of the yolk a few nuclei occur, (YC, Fig. XLV). At what appears to be a later stage, the yolk consists of pyramids with their apices at the centre of the egg. On the bases of the pyramids at the surface of the egg are nuclei or nucleated cells. Within the yolk at this stage, no nuclei were found, though they might have been present and have escaped observation, as the sections were in many cases considerably broken.

Later in development the yolk pyramids break up. This breakage is probably effected by the vesicles in the pyramids uniting and consequently causing a separation of the yolk substance along the line of their union.

These pyramids recall the yolk pyramids of Astacus, described by Reichenbach.[1] They might also be compared to the yolk columns described by Ludwig[2] for the spider's egg. Other investigators, however, have not confirmed Ludwig's observations, but find that all the cells are not at the surface, as he claimed, but that some remain centrally located in the yolk. The yolk of the Mantis egg is very like the yolk of the grasshopper egg. The yolk of the higher orders of insects studied, consisted of rounded spherules which were often much vesiculated.

[1] Die Embryoanlage u. erste Entwicklung d. Flusskrebses. Zeit f. wiss. Zool. Vol. XXIX.
[2] Ueber die Bildung des Blastodermes bei den Spinnen.

The earliest trace of the embryo was obtained from sections of the eggs of the Mantis. At this early stage the embryo is a mass of undifferentiated cells lying on the surface of the egg. (Fig. XLI, E.) The blastoderm is not formed at this stage. Some cells remain in the yolk; whether they are all at the surface at some subsequent stage could not be determined.

Figures XLII & XLIII, show two stages in the formation of the amniotic folds of Mantis. It will be seen that the amnion, as seen in cross sections, arises on each side of the embryo as folds of blastoderm, which meet and unite over its middle line.

When the union of the folds is effected, and the embryo is separated from the surface and covered ventrally by the amnion, the inner layer is formed, as in Meloe and Thyridopteryx, as an ingrowth from the median line of the embryo. Whether the inner layer gives rise to both mesoderm and endoderm could not be determined in the case of Mantis as no advanced embryos of this insect were studied. Figure XLIV is a drawing of a transverse section of an early Mantis embryo, showing the origin of the inner layer [1] from the median groove. From a study of more advanced grasshopper embryos, however, it seems probable that the yolk cells do not take part in the formation of the endoderm; consequently that layer is formed in the grasshopper as in Thyridopteryx from the inner germ layer. Ayers [2] in his memories on Œcanthus, an Orthopterous insect, described undifferentiated cells which never reach the surface to form blastoderm but remain in the centre of the egg as yolk cells. Unlike the corresponding cells of Thyridopteryx, these cells, according to Ayers, take part in the formation of the endoderm. The serosa or outer embryonic membrane is taken into the body cavity of the embryo through its enclosed dorsal surface, and also forms parts of the endoderm. Nothing at all corresponding to such an absorption of embryonic membranes was observed in the advanced grasshopper embryo. This embryo is perhaps best understood by following the series of cross sections represented by figures XLVII–LXV. Most of the sections figured were selected from a single series from an advanced embryo. They were drawn in order, from before backwards. Figure XLVII, represents a cross section of the head of an advanced grasshopper embryo. Reference to the meaning

[2] Ayers. Memoirs of the Boston Society of Natural History, Vol. III no. VIII.

of the letters will explain the details of the figures.   C" in the figure, is the cross commissure of the supra-œsophageal ganglion, crossing above the œsophagus.   At a point on the dorsal surface of this region of the embryo the yolk spherules lose their definite outlines, and appear to run into a homogeneous mass, (YS in Fig.), in which are cells, which are, apparently, migratory mesoderm cells.

Figure XLVIII represents a section of the supra-œsophageal ganglion back of the section just described.   It shows the position of the antennæ and labrum. These correspond in every way to the antennæ and labrum of other insects which have been described.

The amniotic folds of opposite sides have not met in this region.   It will be observed that the amnion at this stage does not cover the ventral surface of the embryo, but has been apparently absorbed at all points except the dorsal extremity of the body wall, where it persists as shown in the figures.

Figure XLIX represents a section, posterior to those previously described, through the anterior part of the œsophagus.   I in the figure represents the inner layer given off from the median line at this point.   It has been carried inwards by the œsophageal invagination.

Figure L is a drawing of a succeeding section in which the circum-œsophageal commissure of the supra-œsophageal ganglion is represented as surrounding the œsophagus.   In succeeding sections this commissure disappears.

In figure LI it will be observed that the mesodermic ingrowth pushed in with the inward growth of the œsophagus forms a hollow thickening on its dorsal surface. This mesodermic thickening may be the rudiment of the heart in the head region, as it apparently corresponds to a similar thickening which forms the head portion of the heart in Thyridopteryx.

Figures LII–LVII represent sections through the mandibular region.   In figures LV, LVI, LVII, it will be noticed that the amniotic folds have met and united on the dorsal surface of the body.   In figure LV the thickening of mesoderm on the dorsal surface of the œsophagus previously referred to as the possible mass from which in the head region the heart originates, is again shown.

In figures LV, LVI, LVII & LVIII, the ganglion, marked No. II, is the mandibular portion of the sub-œsophageal ganglion.   Figures LIX, LX, LXI, represent sections through the first pair of maxillæ. In this region the œsophagus terminates.

The amniotic folds and ectoderm have not covered the dorsal surface of the body, which is, however, closed by mesoderm cells.

The supra-œsophageal ganglion extends through the first maxillary region of the head. It will be noticed that the ganglion is separated from the superficial ectoderm which is particularly thick on its dorsal surface. Possibly the compound eyes are formed from this thickened portion of ectoderm.

The maxillæ have each two lobes outside of and at the base of the main axis of the appendage. (XL. in Fig.)

These lobes recall, though they are probably not homologous with, the exopodite and epipodite of the Crustacean appendage. Similar lobes have been described by Patten for the maxillary appendages of Blatta.

Figure LXII is through the region of the second maxilla. The body wall is here closed dorsally by mesoderm and the amniotic folds extend farther towards the ventral surface than in preceding sections.

Figure LXIII represents a section through the maxillary region of what was probably a somewhat older embryo.

Figure LXIII is a portion of the same section more highly magnified. No. III in the figure represents the ganglion of the second maxilla. I represents the mesoderm which here as elsewhere covers the dorsal surface before the latter is closed by the amnion and ectoderm.

Beneath the dorsal mesoderm lie large cells with granular protoplasm and large deeply granular nuclei. These may be described as blood cells (BC. in Fig.) In the spider somewhat similar cells form the plasma and corpuscles of the blood.

Figures LXIV and LXV represent sections through the thoracic region. These figures hardly require explanation. It will be observed that the amnion, though more extended than in preceeding figures of this series, does not cover the ventral surface of the embryo.

To sum up briefly the observations made on the embryology of the Orthoptera,—

At one stage all the cells are probably on the surface at the bases of the yolk pyramids. Yolk cells must then arise by migration from the surface cells. The yolk cells probably take no part in the formation of the endoderm; for when the body wall was closed dorsally by mesoderm no yolk cells were observed in the body cavity, though they were present in the yolk. The early embryo of Mantis is a mass of undifferentiated cells lying on the surface of the yolk like the primitive cumulus of spiders.

The amniotic folds and inner layer arise as in other insects described. At later stages the amnion is incomplete, not covering the ventral surface of the embryo.

The maxillæ have additional lobes at the base of the main axis of the appendage.

Korotneff [1] has made a study of the embryology of Gryllotalpa by means of sections. An abstract of his work is given here. On comparing it with the results arrived at from the study of the Orthopterous insects, mantis and grasshopper, there appear to be some important points of difference which can hardly be due entirely to the more complete developmental history obtained by Korotneff.

In the earliest stage observed by Korotneff there were four amœbiform cells in the yolk. On the multiplication of these cells and their migration to the surface, the blastoderm is formed. A stage occurs in the formation of the blastoderm during which there are no cells in the yolk. Yolk cells subsequently migrate from the surface; Korotneff states that before the primitive cells. are converted into blastoderm cells, they have for a time no nuclei at all. No such disappearance of nuclei was observed by me in the Orthoptera studied, or in the blastoderm formation of other insects. It might, however, have been overlooked. The granular ill-defined nuclei of the undifferentiated cells appear, from observations made in Thyridopteryx, to become less granular and more vesicular when these cells reach the surface.

The mesoderm of Gryllotalpa, according to Korotneff, arises, not by the separation of a median ingrowth from the outer layer, but by delamination on each side of the median groove. This is certainly not the mode of origin of this layer in mantis. Its origin in different Orthopterous insects may differ, however, for Korotneff's observations seem to have been carefully made. The embryonic membranes of Gryllotalpa arise, as in mantis and other insects described, as folds of blastoderm on each side of the embryo, which meet and unite over its middle line. The united inner limbs of these folds or the true amnion folds rupture in late embryonic life and are absorbed.

[1] Korotneff. Die Embryologie der Gryllotalpa. Zeit. f. wiss. Zool. 1884.

It will be remembered that the amnion, in the late stages of development of the grasshopper, is absorbed, with the exception of a portion attached to the dorsal extremities of the body walls of the embryo.

The outer membrane or serosa, which is equivalent to the original blastoderm minus that portion which forms the amnion, is also absorbed. But before this absorption occurs a layer of yolk cells is formed beneath it. This layer becomes specially thickened on the dorsal surface of the yolk and forms a dorsal organ. A similar dorsal organ has been described for other insects. It is, however, totally different from the dorsal organ described by Brant for Libellula. It will be remembered that no corresponding structure occurs in the development of Thyridopteryx. The dorsal organ becomes concentrated at one point on the dorsal surface. Its cells multiply and migrate into the yolk which they digest or prepare for digestion. The body walls are then closed dorsally; the dorsal organ for a time remains as a tube on the surface of the body. A similar behavior of the dorsal organ has been noticed in other insects. In the latest stages which I studied, it will be remembered that in the grasshopper the serosa or blastoderm persisted on the surface of the yolk; and there appeared to be nothing present or in process of formation which corresponded to a dorsal organ.

According to Korotneff, the mesoderm is distinctly divided into splanchnic and somatic portions The former enclosing in the body cavity the yolk which is arranged in pyramids with yolk cells at their bases. No marked separation of mesodermic layers was observed in the grasshopper.

Migratory mesoderm or mesenchyme cells arise from the mesoderm. These resemble similar cells described in Thyridopteryx. Blood cells arise from mesoderm on the median ventral surface. Their histological structure is not described by Korotneff. They may, however, be like the cells described as blood cells in the advanced grasshopper embryo.

The heart is formed from folds of the somatic layer of mesoderm growing together over the dorsal surface somewhat in the manner described by Dohrn. [1] The nervous system arises according to Korotneff as a single string which becomes sep-

[1] Notzien zur Kenntniss d. Insectenentwicklung.

arated by a median depression, which is formed where the first groove or primitive furrow had previously occurred.

The cells forming the floor of the depression separating the nerve strings take part in the formation of the nervous substance. This does not agree with my observations on the origin of the nervous system in Thyridopteryx, where the median groove did not disappear before the formation of the nervous system. The latter did not appear first as a single string. The cells lining the median groove separating the nerve strings did not apparently form any of the nervous substance.

DIPTERA.

Before going to the embryology of spiders, it may be well to insert here some observations made on the maturation of the egg in Musca (domestica?).

Each ovarian tube of the fly's ovum is divided, at certain stages at least, into five divisions, viz: the end chamber (EC Fig. LXVI,) and four following chambers, the largest (CB, IV, Fig. LXVI.) at the end of the tube.

The ovarian tube consists externally of a membraneous peritoneal sheath in which nuclei are embedded (FE in Fig. LXVI.)

The end chamber contains nuclei, the histological structure of which could not be determined.

Korschelt, from observations on Musca vomitaria, finds that the egg cells and nutritive cells which form the contents of succeeding chambers arise from the larger nuclei contained in the end chamber, while the smaller nuclei of the end chamber assume a superficial position in succeeding chambers, enclose the germinal vesicle and nutritive cells and become the nuclei of the ovarian epithelium.

Will, [1] in an elaborate article on the origin of the yolk in insects, claims that the nuclei of the end chamber (the so-called ooblasts) give off portions of their substance, which form in some cases the nuclei of the ovarian epithelium and in others the nuclei of the nutritive cells.

The remains of the nucleus of the ooblast then form the germinal vesicle and spot.

From the incomplete observations made on the embryology of Musca, it seems that Korschelt's view is the correct one. There does not appear to be any budding

[1] Will. Zeit f. wiss Zool. 1885.

from the nucleus of the ooblast to form ovarian epithelium and nutritive cells; on the contrary, the nuclei after leaving the end chamber are arranged in circular masses, the outer nuclei of which are smaller than the enclosed nuclei. The latter, in more advanced egg chambers, can be recognized as the nutritive cells, while one differing from the nutritive cells in appearance becomes the germinal vesicle.

The germinal vesicle was not observed in the younger egg chambers, that is, those nearest the end chamber of the tube. Each chamber of the tube with the exception of the end chamber forms a single mature egg.

Figure LXVII represents the last chamber of the egg tube in a more advanced stage than that represented by CBIV Fig. LXVI.

The ovarian epithelium (OVE in Fig) has become quite columnar at the end of the egg chamber towards the outer end of the ovarian tube. The nutritive cells at this end of the chamber have broken down, forming yolk, (Y in Fig. LXVII). The yolk consists of small spherules, in every respect similar to spherules found in the nuclei of the nutritive cells (N, C in Fig.) The yolk of the mature fly's egg is of a similar character. In all probability then the yolk consists of the broken down nuclei of the nutritive cells. In the mature ovum it is surrounded by a protoplasmic sheath which is probably derived from the protoplasm separating the nuclei of the nutritive cells.

Figure LXVIII represents a longitudinal section of a more advanced stage in which the germinal vesicle is seen lying in the yolk. It consists of finely granular protoplasm which, unlike the nuclei of the nutritive cells, does not stain. At this stage, the vesicle has a definite boundary. It is eccentric in position, lying near the ovarian epithelium, which at this stage, has begun to excrete the chorion (CH in Fig.) on its inner surface. The ovarian epithelium apparently takes no part in the formation of the yolk. After the formation of the chorion it adheres as flattened hexagonal cells to the outer surface of the latter even when the egg is laid.

Figure LXIX is a drawing of a cross section through an egg chamber at the same stage of development as that represented in figure LXVIII. In this figure the deeply stained germinal spot is represented in the centre of the germinal vesicle (GS in Fig).

The germinal spot is not homogeneous throughout, but has in its centre what appears to be a swelling with a depression on its apex. However that may be, the central portion of the spot has a different refractive index from the peripheral portions.

Figures LXIX and LXIX' represent a later stage in the maturation of the egg drawn with a low and a high power.

The germinal vesicle at this stage has lost its definite boundary and shades off into the yolk spherules.

The ovarian epithelium has grown round and nearly closed the outer pole of the egg.

Through the opening still left by the ovarian epithelium the remaining unabsorved nutritive cells are probably taken in. The germinal spot, at this stage, differs from the same structure in earlier stages. This, however, may be due to the different action of hardening fluids or to differences in preservation, etc.

Sections of the mature egg showed no traces of the germinal vesicle or spot. The latter, as we may believe from Hertwig's [1] observations on the maturation of the ovum, might readily persist and yet be overlooked, owing to its small size and to the confusion resulting from the number of yolk spherules. Will [2] in the article referred to describes the migration of the vesicle of this egg to the periphery where it loses it boundary and runs into the yolk. He thinks the germinal spot also disappears.

ARACHNIDA.

The observations made on the embryology of spiders were not complete, but they have brought to light some interesting points not heretofore noticed or sufficiently emphasized. No observations were made on the early stages of segmentation. The observations of Ludwig on the early stages of development do not, as before stated, fully accord with those of other observers. Schimkewitsch agrees with Balfour in thinking that on the formation of the blastoderm some undifferentiated cells remain in the yolk cells. The observations recorded here do not settle this point however.

Before the formation of the blastoderm, the so-called primitive cumulus is formed on the surface of the egg.

At this stage there are cells undergoing division in the yolk (YC Fig. LXXV).

[1] Hertwig. Morphologishes Jahrbuch Vol. III.

Balfour says the cumulus appears after the blastoderm has been fully formed. This is not true of the species of spider studied by me. The primitive cumulus consists of a mass of undifferentiated cells, extending well into the centre of the egg. (Fig. LXXV.) Their histological structure is well preserved in the section which figure LXXV represents. The cells have large granular nuclei. The protoplasm is marked by granular radii extending from the nucleus to the periphery of the cell. These cells are primitive and unspecialized. They resemble the yolk cells which are certainly like the cells occurring in the early stages of segmentation before specialization of tissue has taken place. It is not important to enquire whether the cumulus results from the division of surface cells or is formed by the accession of cells from the yolk. It is very probably formed both by the division of cells which have reached the surface, and by the addition to these of yolk cells. It is important to note, however, that it consists of undifferentiated cells, and is formed, in the species of spider which I studied, before the blastoderm is completed. In Limulus the first trace of the embryo is also a mass of undifferentiated cells lying at the surface of the egg.[1]

Figure LXXVI represents the next stage obtained after the cumulus stage. It is a longitudinal section showing nine or ten mesoblastic somites which are not hollow at this stage.

It seems that the cumulus must take part in the formation of some if not all of these somites.

In Limulus the mesoderm arises in part at least from the inner cells of the cumulus.

Balfour states that the somatic mesoderm arises from the ectoderm and the splanchnic mesoderm from the yolk cells. In the early stages, however, the mesoderm is not divided into two layers. Figure LXXVII represents yolk cells near the surface undergoing what appears to be endogenous division. This corresponds to an endogenous division observed by Reichenbach as taking place in the yolk cells of Astacus.

Figure LXXVIII represents a longitudinal section of an embryo more advanced than that represented by figure LXXVII. Here the six thoracic and four provisional abdominal appendages have appeared.

[1] University Circulars, 1885.

The mesodermic somites are hollow, the somatic portion lining the cavities of the appendages, the splanchnic portion not entering the appendages. At this stage there is posterior to the last thoracic appendage a swelling marked OP in figure LXXVIII. Although this was not observed in all cases, it is apparently a normal structure, for it is present in later stages. It corresponds in position to the operculum of Limulus. Figure LXXIX is a drawing of a longitudinal section of a still more advanced embryo. The first two abdominal appendages (A & B in Fig.) are not well marked but the mesoblastic somites corresponding to them are well defined (A' & B' in Fig.) Figure LXXIX represents a more highly magnified portion of the same section.

The two abdominal appendages, A & B, are better defined in reality than in the figure.

It will be seen that the ectoderm covering the two appendages, A & B, is columnar, each cell having a well defined chitinous boundary and a square nucleus.

These cells correspond, in histological structure, to those forming the laminæ of the lung book of the adult spider as described by McLeod.[1] The histological structure of the cells of the gill lamellæ of Limulus is quite similar.

On the anterior surface of the appendage, A, will be seen a fold (I in figure.) If we imagine the appendage to be pushed further in and come to lie entirely in the lung cavity L, B, the fold on the anterior face of the appendage and others which may arise there will then lie on the anterior wall of the lung cavity, but these folds will then be directed backwards and not forwards as before the involution of the appendage and will consequently correspond in every way to the laminæ of the lung book. All the stages in the involution of the appendage were not traced, but there can be little doubt that the lung book of the spider results from the involution of embryonic abdominal appendages. Such an involution of appendages to form the lung book has been suggested by Lankester[2] on theoretical grounds, but not from observation. The involuted appendages are covered by the structure described as the operculum. (OP, Fig. LXXIX'.)

Some remarks on the supra-œsophageal ganglion and structures connected with it will conclude the observations made on the embryology of spiders.

(1) Archiv. de Biologie, 1884.
(2) Quarterly Journal 1885.

The supra-œsophageal ganglion was studied by transverse and longitudinal sections.   Figures LXXI, LXXII, LXXIII, LXIV, represent transverse sections of the brain of an advanced embryo cut backwards from the anterior part of the œsophageal invagination.   On each side of the brain will be seen folds (AM in figure) which correspond to the amniotic folds of insects.   In the last transverse section (Fig. LXXIV,) the outer limbs of the folds have separated from the inner limbs and have united.   The thicker inner limbs of these folds have not united but in other respects they correspond completely to the true amnion of the insect embryo.

It will be seen from a study of longitudinal sections (Figures LXXX, LXXX' LXXXI, LXXXII', LXXII) that these folds occur in the head region only, and cover but a part of the supra–œsophageal ganglion.

Figure LXXX, is a drawing of a longitudinal section of an advanced embryo laterad of the median line.   Figure LXXX' represents the cephalic portion of the same section drawn with a higher power.   AM in the figure represents the so-called amniotic folds in cross section.   No. I is the anterior portion of the supra-œsophageal ganglion.   No. I' is the posterior division of the same ganglion.   MD represents the mandible.   The portion of the brain above it represents the mandibular division of the supra-œsophageal ganglion.

The large cells (B, C, in the figure) with granular protoplasm and well defined nuclei may be termed blood cells.   On the fusion of these cells their protoplasm appears to form blood plasma; and their nuclei, with an investment of protoplasm in some cases, the blood corpuscles

Figure LXXXI represents a section of the same series farther towards the median line of the embryo.

LXXXI' is a drawing of the same section more highly magnified.

Here it will be seen that the inner amniotic fold has separated from the outer and adheres closely to the anterior portion of the supra-œsophageal ganglion.   The outer limb of the amnion fold extends over the supra-œsophageal ganglion to its mandibular division.

The blood cells are kept from passing out between the amnion folds by mesodermic strings extending from the mesoderm which invests the supra-œsophageal ganglion to the outer amnion fold.

Figure LXXXII represents a section of the same series close to the median line. Here it will be observed that the inner amniotic fold has separated from the outer and lies close to the supra-œsophageal ganglion, to which it is apparently attached. Blood cells have broken down between the two folds and have thus formed blood plasma and blood corpuscles. Blood cells are numerous in the dorsal surface of the embryo, where they generally lie between the two layers of mesoderm, indicating perhaps that the blood once circulated through the body cavity. Balfour describes the groove regarded here as amnion, as a depression on the anterior and lower surface of the procephalic lobes. Metschinkoff,[1] in his article on the embryology of scorpions, has described a fold which closes in the ventral surface of the brain of the scorpion, and is probably similar to the amnion of spiders. If this fold were extended the whole length of the embryo it would correspond in every way to the amnion of insects.

The observations presented here on the embryology of spiders can be very briefly summarized.

The primitive cumulus, consisting of undifferentiated cells, appears before the blastoderm is fully formed.

It probably forms a considerable part of the mesoderm. According to Balfour it occupies a position where the caudal lobe of the embryo subsequently appears.

The mesoblast, or part of it at least, must then grow forward from this posterior part of the embryo.

Amniotic folds appear in the head region of the embryo.

In the species of spider studied, probably two abdominal appendages are invaginated to form each lung book.

The supra-œsophageal ganglion is indistinctly divided into two portions.

CONCLUSION.

It may be well to close this paper with some remarks on the relations of tracheates suggested in part by the observations given.

[1] Embryologie des Scorpions. Zeit. f. wiss. Zool. Bd. XXI.

Peripatus and Myriapods, from the absence of wings and other primitive characters, may fairly be considered the most primitive tracheates.

The position of Peripatus is uncertain ; but some Myriapods show indications of a hexapod stage in their development. They may therefore be related to the wingless Hexapods. The large number of body segments and appendages in Myriapods, and perhaps in Peripatus as well, is probably only the vegetative reproduction of homologous parts.

In Myriapods and in Peripatus as shown by the studies of Kennel [1] and Sedgwick [2] on the latter, and by observations of other observers on the former, the segmentation is total. From the accounts given by Sedgwick and Kennel it appears that the gastrulation differs in different species of Peripatus.

The mode of origin of the endoderm is not, however, very important for classificatory purposes, inasmuch as it is very likely to be modified by the presence or absence of food yolk.

The mesoderm, in the development of Peripatus, grows forwards from an undifferentiated cell mass at the posterior end of the embryo. The mesoderm arising from the primitive cumulus of spiders also grows forwards from an undifferentiated cell mass at the posterior end of the embryo.

The endoderm of Peripatus, Myriapods and Spiders is derived from the inner layer of the gastrula. How the inner layer of the gastrula arises is unimportant. Consequently Peripatus and spiders are quite alike in the formation of germinal layers.

This alone, however, does not indicate any close relationship, for in Crustacea the endoderm arises from the inner layer of the gastrula while the mesoderm grows forward from the posterior end of the embryo as in Peripatus and spiders.

In the higher insects the yolk cells, from their mode of origin, probably represent the inner layer of the gastrula and are consequently equivalent to the endoderm of lower forms. The true endoderm is functional only during embryonic life in absorbing the yolk. It takes little or no part in the formation of the digestive tract. In these higher insects, as already shown, the inner layer, which from its

[1] Kennel. Entwicklungsgeschichte der Peripatus Edwardsii. Sempers' Arbeiten 1884.

[2] Sedgwick, Development of Peripatus Capensis. Quarterly Journal, 1885.

position and segmentation corresponds to the mesoblast of Arachnids and Peripatus, has usurped the functions of the true endoderm.

In Aphides from the studies of Witlaczil it appears that the intestine is formed exclusively from the proctodæal and stomadæal invaginations.

In order to separate the different divisions of the arthropod phylum anatomical characters as well as embryological phases must be considered.

The posession of a single well developed pair of antennæ, of tracheal invaginations and of embryonic membranes, and the existence of a hexapod stage in their development afford sufficient ground for regarding myriapods as lowly organized or degenerate insects. Peripatus would perhaps come under the same category through the embryonic membranes of Peripatus do not appear to correspond fully to those of insects.

Arachnids, from the absence of antennæ and the histological structure of the abdominal appendages, and from other characters, anatomical and histological, must certainly be included, with Limulus, in a distinct group of arthropods. The small seventh pair of thoracic appendages of Limulus is perhaps an interpolated appendage, or perhaps a corresponding appendage may be discovered in the development of spiders and scorpions.

Arachnids, probably, never possessed antennæ, since all their appendages, like those of Limulus, are at one period post oral, and are not innervated by the supra-œsophageal ganglion. Trilobites, possibly the ancestrial form of Limulus, from evidence afforded by the Cincinnati specimen, probably possessed no antennæ. If antennæ were ever present in the group we would expect to find them in these old forms.

The antennæ of insects from their innervation correspond to the first pair of crustacean antennæ. The bilobed upper lip of insects is innervated from the second division of the supra-œsophageal ganglion which forms part of the circum-œsophageal commissure. In the nauplius stage the second pair of crustacean antennæ is innervated from the circum-œsophageal commissure. From their similar innervation a comparison may then be fairly drawn between the paired upper lip of insects and the second pair of crustacean antennæ.

The antennæ of insects and crustacea are probably homologous structures and ally the two groups.

The amnion of insects and arachnids is probably homologous, and allies these two groups; consequently insects, having characters common to both arachnids and crustacea, may be placed between arachnids on the one side and crustacea on the other. It does not follow from this that insects, from the possession of common characters, are the most primitive form, but rather that they are the most heterogeneous and highly evolved of the arthropod groups.

The three arthropod groups may not have arisen one from the other; probably each group arose independently from a common source.

The tracheæ of insects and spiders are probably analogous, not homologous structures.

This may fairly be concluded from the fact that the tracheæ of spiders can be derived from the lung books which it will be remembered are involuted appendages. Consequently we could not expect to find in insects tracheal invaginations occurring in appendage-bearing segments of the body if the tracheæ of the two groups were homologous structures. It is needless to say that we do find tracheal invaginations in insects occurring in appendage-bearing segments. The tracheæ of insects and spiders are therefore not homologous.

EXPLANATION OF PLATES.

REFERENCE LETTERS.

Am.—Amnion.

Am. F.—Amniotic folds.

Am.—Portion of amnion folds which forms dorsal surface of body.

App.—Thoracic appendages.

BL.—Blastoderm.

BH.—Blastopore.

C".—Sub-œsophageal commissure of supra-œsophageal ganglion.

C'.—Supra-œsophageal commissure of supra-œsophageal ganglion.

C.—Circum-œsophageal commissure.

CB.—Chambers of ovarian tube of Fly.

CH.—Chorion.

E.—Embryo.

E.'—Vesiculated portion of yolk.

EC.—Ectoderm.

FE.—Folicular epithelium.

GL.—Salivary gland.

GV.—Germinal vesicle

GS.—Germinal spot.

H.—Heart.

I.—Inner layer.

I'.—Migratory cells of inner layer.

IS.—Somites of inner layer.

Ie.—Endodermic portion of inner layer.

LB.—Labrum.

LB. N.—Labial nerve.

M.—Muscle.

MD.—Mandible.

MX.—Maxilla.

MX'.—Second maxilla.

MXL.—Maxillary lobes.

NS.—Nervous system.

NC.—Nutritive cells.

OVE.—Ovarian epithelium.

PC.—Procephalic lobes.

S.—Cuticle.

YC.—Yolk cell.

YB.—Yolk ball.

YS.—Yolk spherule.

YS'.—Homogeneous yolk mass.

No. I.—Anterior portion of supra-œsophageal ganglion.

No. I'.—Posterior portion of the same.

No. II.—Mandibular portion of sub-œsophageal ganglion.

PLATE I.

Fig. I. Transverse section of ovarian egg of Thyridopteryx. OE—Ovarian epithelium. YS—Yolk spherules. P—Protoplasmic stratum surrounding yolk. CH—Chorion.

Fig. I'. Portion of same section more highly magnified.

Fig. II. Segmenting egg of Thyridopteryx. VC—Yolk cells most of which ultimately reach the surface.

Fig. III. Transverse section of egg of Thyridopteryx with blastoderm partially formed about yolk. Yolk spherules are not represented.

Fig. III'. Yolk cells at this stage reaching the surface of the egg.

Fig. IV. Transverse section of egg of Thyridopteryx showing completed blastoderm (BL); early embryo (E), and commencing amniotic folds (AMF).

Fig. IV'. Highly magnified embryonic area of same section.

Fig. V. Transverse section of egg of Thyridopteryx showing embryo, (E) in cross section situated in centre of yolk. Amnion covering its ventral surface is not represented. The yolk is segmented.

Fig. VI. Hexagonal blastoderm cells.

Fig. VII. Transverse section of embryo of Thyridopteryx showing formation of inner layer (I).

Fig. VIII. Horizontal section of embryo of Thyridopteryx showing procephalic lobes (PC) at anterior end.

Fig. VIII'. Longitudinal section of embryo at same stage showing segmentation of inner layer (IS).

Fig. IX. Horizontal section of embryo at same stage as that represented in Fig. VIII showing its relations to blastoderm BL.

Fig. X. Transverse section of Thyridopteryx embryo showing separation of inner layer (L) beneath the blastopore BH.

Fig. X'. Same section less highly magnified showing its relations to blastoderm (BL).

Fig. XI. Transverse section of older embryo showing origin of nervous system (NS).

Fig. XII. Nervous system separated from outer ectoderm. Elongated cells at bottom of median groove.

Fig. XIII. Transverse sections through region of procephalic lobes.

Fig. XIV. Transverse section through same region of older embryo. Inner cells of lateral portions specialized as nerve cells.

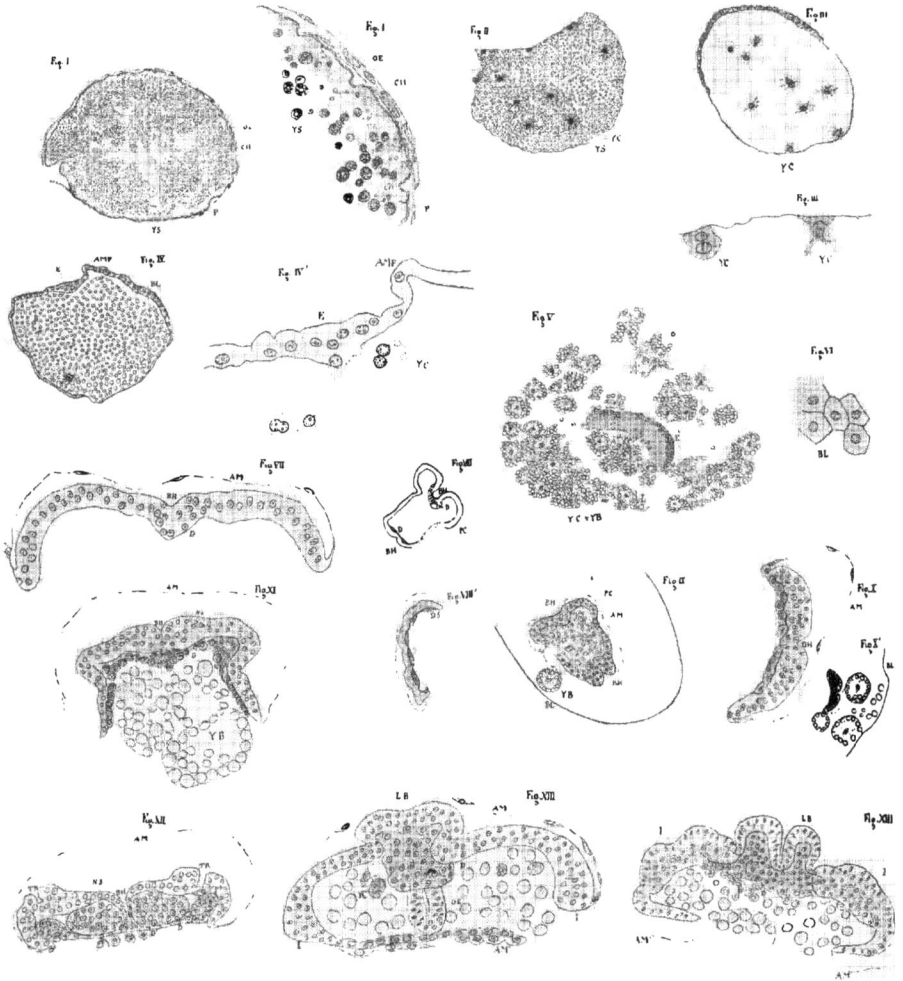

BRUCE. Insects and Arnchnids

PLATE II.

Fig. XV. Transverse section of advanced Thyridopteryx embryo. Nervous system (NS) separated from the ectoderm. Amnion (AM) growing dorsally. Portion of the inner layer (IE) becoming specialized as endoderm.

Figs. XVI–XXI. Longitudinal sections of Thyridopteryx embryo.

Fig. XVII. Is nearly median. Numbers 1–17 indicate somites which have ganglia corresponding to them.

Fig. XXII. Cephalic portion of XVII highly magnified.

Fig. XXIII. Last two thoracic (Nos. VI, VII) and first two abdominal ganglia (Nos. VIII, IX) of figure XVIII, highly magnified.

Fig. XXIV. Transverse section of advanced Thyridopteryx embryo showing approximation of amnion folds (AM) on dorsal side of embryo. Endodermic portion of inner layer (IE) beginning to shut yolk into body cavity. Migratory mesoderm (I') growing round yolk before the endoderm.

Figs. XXV, XXVI. Transverse sections of same embryo, to show the approximation of amniotic folds on dorsal side of embryo.

Figs. XXVII–XXVIII. Transverse sections of the head of an advanced Thyridopteryx embryo. The plane of the section is approximately an imaginary line joining numbers 1 and 2 in Fig. XVII.

Fig. XXIX. A diagramatic representation of the under surface of the head of an advanced embryo.

Fig. XXX'. A diagramatic lateral view of the head of an embryo at a corresponding stage of development.

PLATE III.

Figs. **XXIX, XXX, XXXI.** Transverse sections of the head of an advanced embryo They are from the same series as figures XXVII, XXVIII.

Fig. **XXXII.** A transverse section of the same series as the preceeding, through the posterior part of sub-œsophageal ganglion (No. II).

Fig. **XXXIII.** Transverse section of head of Chrysopa showing relation of last abdominal somite to preceeding abdominal somites.

Fig. **XXXV.** Transverse section of egg of Melöe showing formation of blastoderm.

Fig. **XXXV'.** Structure of blastoderm.

Fig. **XXXVI.** Transverse section of the egg showing origin of inner layer from median groove (BH), and amniotic folds (AMF).

Fig. **XXXVII.** Transverse section of a more advanced embryo of Melöe. The amniotic folds have united and the embryo has consequently been separated from the surface.

Fig. **XXXVIII.** Transverse section of embryo at same stage but through a different region.

Fig. **XXXIX** Transverse section of more advanced embryo of Melöe in which inner layer (I) has been constricted off

Fig. **XL.** Transverse section of advanced embryo of Melöe The nervous system (NS) has been separated from the surface.

Fig. **XLI.** Transverse section of egg of Mantis showing embryonic area E.

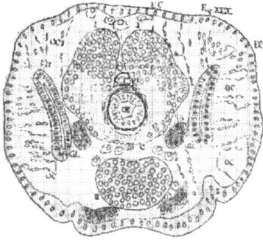

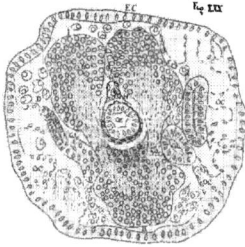

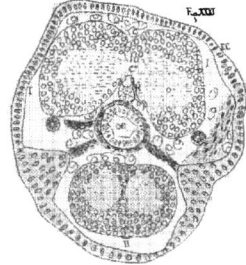

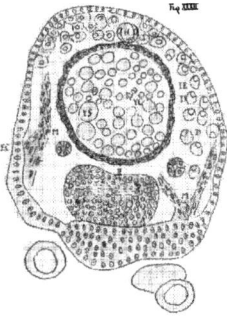

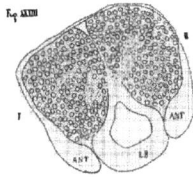

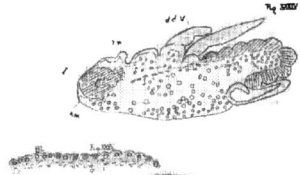

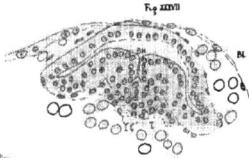

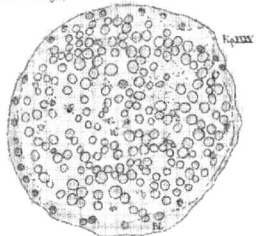

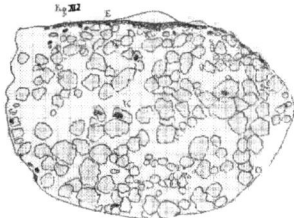

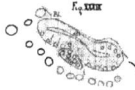

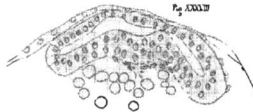

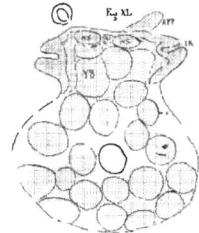

PLATE IV.

Fig. XLII. Transverse section of early Mantis embryo (E) with amniotic folds (AM) on each side

Fig. XLIII. Transverse section of more advanced Mantis embryo almost separated from surface by union of amniotic folds.

XLIV. Transverse section of still older Mantis embryo showing origin of inner layer (I).

Fig. XLV. Transverse section of early stage of segmentation of grasshopper egg, showing a few cells (YC) in yolk.

Fig. XLVI. Transverse section of a later stage of grasshopper egg showing yolk pyramids with cells at their bases.

Fig. XLVII. Transverse section of head of advanced grasshopper embryo showing commissure (C) of supra-œsophageal ganglion.

Fig. XLVIII Succeeding section of same series showing antennæ (ANT) and labrum (LB).

Fig. LXIX. Transverse section through commencing œsophageal ingrowth.

Figs. L.-LXII. Succeeding transverse sections of same embryo.

Figs. LII-LVIII. Sections through the mandibular region.

.

PLATE V.

Figs. LXI–LXII. Sections through the first and second maxilla.

Fig. LXIII. Transverse section of an older embryo through the first maxillæ. It is closed dorsally by mesoderm; beneath this lie the blood cells.

Figs. LXIV–LXV. Transverse sections of the thoracic region Body wall completed dorsally by mesoderm; the amnion is incomplete.

Fig. LXVI. Longitudinal section of ovarian tube of fly; CB'–CB^{IV} represent sections of successive chambers of the tube.

Fig. LXVII. Longitudinal section of a more advanced terminal chamber (C^{IV}).

Fig. LXVIII. Longitudinal section of a more advanced terminal chamber (C^{IV}) showing elongated epithelium (OVE) excreting chorion (CH) and germinal vesicle (GV).

Fig. LXIX. Transverse section of terminal chamber at same stage as preceeding through germinal vesicle (CV) and germinal spot (CS).

Fig. LXX. A longitudinal section of terminal chamber showing germinal vesicle (GS) without boundary.

Fig. LXX'. Portion of same section more highly magnified.

Figs. LXXI–LXXIV. Sections through the œsophageal region of a spider embryo showing amniotic folds (AM) on each side of the supra-œsophageal ganglion, the outer limbs of which in the most posterior section (Fig. LXXIV) have united.

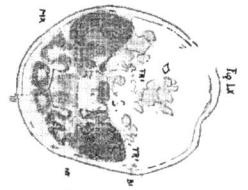

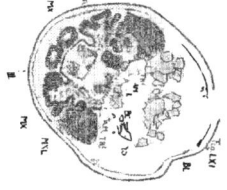

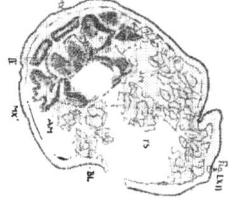

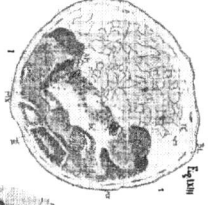

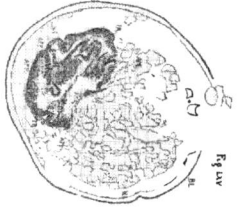

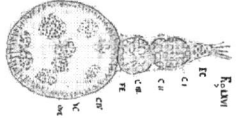

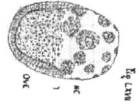

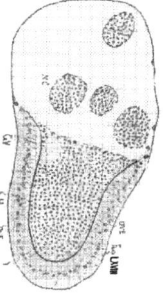

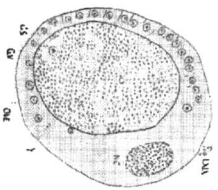

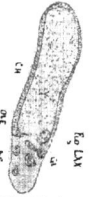

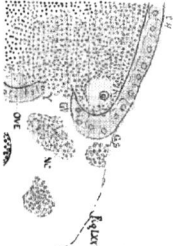

PLATE VI.

Fig. LXXV. Transverse section of spider's egg through primitive cumulus (C).

Fig. LXXV. Cumulus of same section highly magnified.

Fig. LXXVI. Longitudinal section of spider embryo showing solid mesoblastic somites.

Fig. LXXVII. Yolk cells (YC) of advanced embryo apparently undergoing endogenous division.

Fig. LXXVIII'. Longitudinal section of advanced spider embryo with appendages and hollow mesoblastic somites.

Fig. LXXIX. Longitudinal section of more advanced embryo. First two abdominal appendages (A & B) are less distinct, and are being folded into form lung book.

Fig. LXXIX'. Abdominal portion of same figure magnified.

Fig. LXXX. Longitudinal section of advanced spider embryo showing amniotic folds (AM) of brain.

Fig. LXXX'. Cephalic portion of same section highly magnified.

Fig. LXXXI. Longitudinal section of same embryo nearer median line.

Fig. LXXXI'. Cephalic portion of the same highly magnified.

Fig. LXXXII. Highly magnified longitudinal section of brain near median line,

Fig. LXXII

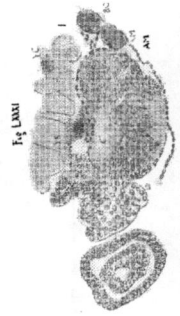

Fig. LXXXI

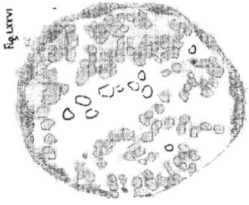

Fig. LXXVI

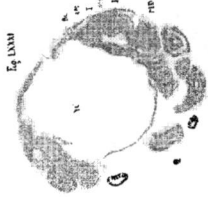

Fig. LXXVII

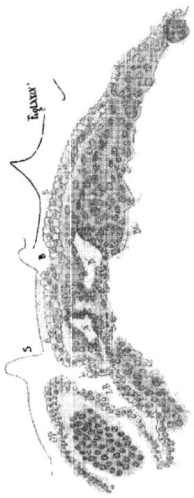

Fig. LXXIX

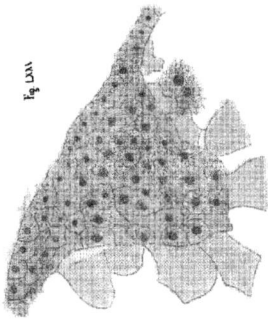

Fig. LXXIII

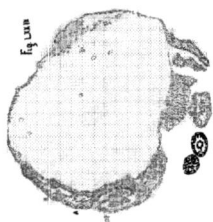

Fig. LXXVIII

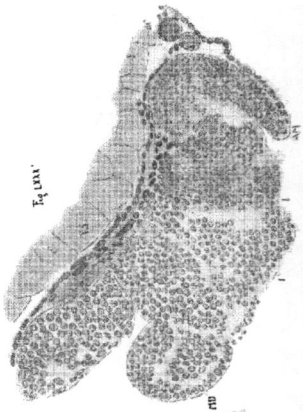

Fig. LXXV

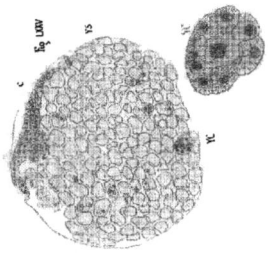

Fig. LXXIV

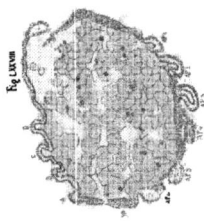

Fig. LXXX

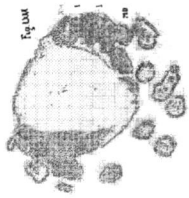

Fig. LXXII

Fig. LXXXII

www.ingramcontent.com/pod-product-compliance
Lightning Source LLC
Chambersburg PA
CBHW022001190326
41519CB00010B/1354